I Wonder
If Sea Cows Give Milk

I Wonder
If Sea Cows Give Milk

and other neat facts about unusual animals

By Annabelle Donati
Illustrated by Paul Mirocha

A GOLDEN BOOK • NEW YORK
Western Publishing Company, Inc.
Racine, Wisconsin 53404

Produced by Graymont Enterprises, Inc., Norfolk, Connecticut
Producer: *Ruth Lerner Perle*
Design: *Michele Italiano-Perla*
Editorial consultant: *Penny Kalk*, New York Zoological Society, Bronx, N.Y.

© 1993 Graymont Enterprises, Inc. All rights reserved. Printed in the U.S.A. No part of this book may be copied or reproduced in any form without written permission from the publisher. All trademarks are the property of Western Publishing Company, Inc. Library of Congress Catalog Card Number: 92-74049.
ISBN: 0-307-11327-2/ISBN: 0-307-61327-5 (lib. bdg.)   A   MCMXCIII

Contents

What in the world is a platypus? 4

Why does the cassowary wear a crash helmet? 6

Do sea cows give milk? 8

Why do hoatzin birds have claws on their wings? 10

Why are mandrills so colorful? 12

Why is the gavial's snout so long? 14

Why do some moths have "eyes" on their wings? 15

Do poison-arrow frogs shoot arrows? 16

Is the okapi a donkey or a zebra? 18

Where are the horseshoe crab's eyes? 20

Does the star-nosed mole's nose shine? 21

Does the devilfish have horns? 22

Why does the hornbill sometimes hide in a hole? 24

Are armadillos reptiles? 26

How does the warthog use its tusks? 28

Does the matamata have worms on its head? 29

Does the wrasse doctorfish make house calls? 30

Tell Me More 32

Index 32

What in the world is a platypus?

Of all the unusual animals on earth, the duck-billed platypus is surely the most unusual. The platypus is considered a mammal because it has fur on its body and the female produces milk for her babies. But this strange two-foot-long creature looks like a weird jumble of many different kinds of animals. It has a flat tail like a beaver's and thick, short fur like an otter's. Like a duck, it has a broad, flat bill, and the female lays eggs. But instead of having a hard shell, the eggs are soft and leathery, like an alligator's. Finally, the male platypus has hollow spurs on his back feet that release a poison similar to a snake's venom.

Where does the platypus live?

The platypus is found only in Australia, where it makes its home in or near streams, rivers, and ponds. Like many creatures that live in water, the platypus has special webbing on its feet that helps it to steer and paddle. But unlike most water animals, it has claws on its feet for digging and for gripping the ground. To make walking easier, the webs on its front feet fold back when the platypus is on land.

Tell Me More

The platypus's eyes and ears are tucked inside pockets that close automatically underwater. Therefore, the platypus can neither see nor hear when it is swimming—which is most of the time. But the animal's bill is so sensitive that it can feel the faint electrical vibrations given off by shrimps and worms, its food, on the river bottom.

How are platypus babies born?

The mother platypus deposits one or two eggs into a deep hole inside a tunnel or burrow that may be entered by land or water. She curls her tail around the eggs to keep them warm. In about ten days, the naked and blind babies hatch. The babies drink their mother's milk until they can find their own food.

Amazing but TRUE

About two hundred years ago, a British naturalist in Australia saw a platypus for the first time. Because the animal was so strange-looking, he sent a platypus skin to some friends in England who were scientists. When the scientists opened the package, they thought that their friend had sewn together the skins of different animals as a joke!

Why does the cassowary wear a crash helmet.

The human-sized cassowary has a bony, helmet-shaped crest that protects its head like a crash helmet. This crest can be used to push through the thick jungle of New Guinea and Australia. But the cassowary's "crash helmet" is only one of this bird's remarkable features.

Amazing but TRUE

The female cassowary lays the eggs, but the male is the one that sits on them until they hatch. He stays with the striped chicks for at least a year.

What makes the cassowary so unusual?

The cassowary's many strange ways make it very different from other birds.

• Sound off
Instead of singing sweetly, as most birds do, the cassowary rumbles, booms, roars, and hisses.

• Grounded
The cassowary cannot fly. It escapes from danger by swimming or running.

• Funny females
Most female birds are smaller and less colorful than their mates. But the female cassowary is larger and as brightly colored as the male.

• Terrible toes
The cassowary is usually quite peaceful and shy, but it can be dangerous when attacked. Its feet are equipped with deadly, razor-sharp claws that can cut an enemy to shreds.

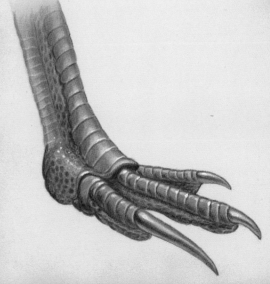

Do sea cows give milk?

Long ago, sailors told stories about strange and wonderful creatures they saw playing in the waters around West Africa, Brazil, and Florida. The sailors were sure that these gentle, playful creatures were mermaids—half fish and half human. Today we think that what those sailors saw were not mermaids at all but manatees, which are sometimes called sea cows.

The only true sea cow, Steller's sea cow, was hunted until there were no more left on earth. This gentle animal has been gone for more than two hundred years. The manatee is smaller and lighter than Steller's sea cow, but in many ways it behaves just as its larger relative did.

How are manatees like cows?

Like dairy cows, manatees and their Pacific Ocean cousins, the dugongs, spend most of the day grazing in fields of grass—underwater grass, that is. These pudgy animals eat up to a hundred pounds of water plants each day. Although farmers don't milk them, female manatees—like all mammals—do produce milk for their babies.

Tell Me More

Most mammals live at least part of their life on land. But manatees—like porpoises, whales, and dolphins—are *aquatic* mammals. They never leave the water. However, manatees are different from other aquatic mammals in that they eat only plants.

How is the manatee like a fish?

Although the manatee is not "half fish," it has much in common with fish. Its body is shaped like a fish's, which helps the animal move easily through the water. Instead of forelegs, the sea cow has flippers that help it to steer and turn; and instead of hind legs, it has a broad, flat tail.

Mother love

Sea cow mothers seem almost human in the way they care for their babies. The young calf swims under its mother's big body, and she shelters and protects the baby with her flippers. Like a human mother, the manatee mother nurses her babies for a much longer time than most other mammals do.

How is the manatee like a person?

Although it is not "half human," the manatee uses its broad flippers in ways that people use their arms and hands. For example, a manatee can pull itself up by its flippers to reach for plants at the water's edge; it can bring food to its mouth and push it in; and it can clean its upper lip and whiskers and wipe its mouth and gums. When it has an itch, it scratches its face or chest with the fingernails at the end of its flippers. And, like people, sea cows sometimes press their muzzles together and "kiss."

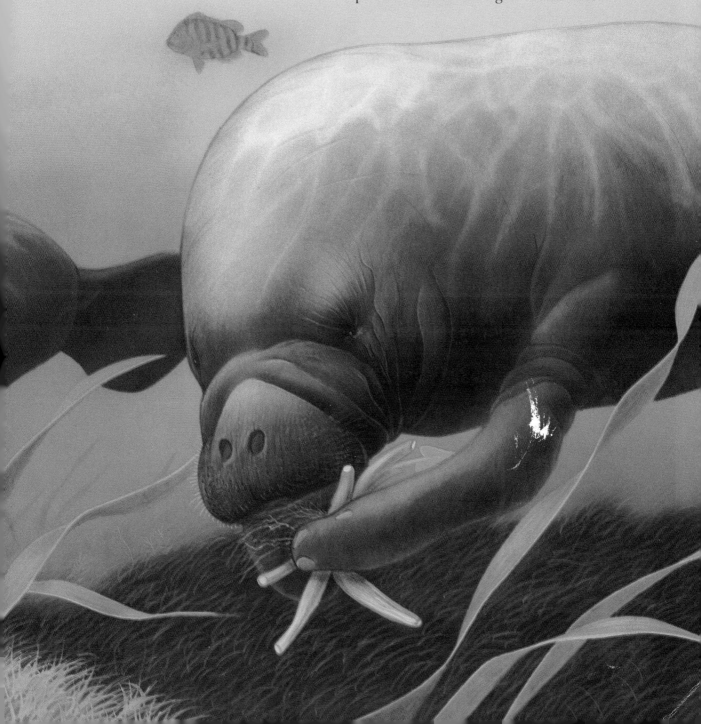

Why do hoatzin birds have claws on their wings?

Newly hatched hoatzin chicks are equipped with claws on their "elbows." Since they have no feathers and can't fly until they are about two months old, they use their wing claws to help them climb trees and walk along branches. As the chicks get older, their feathers start to grow, and the claws on their wings become hidden and are no longer used.

The adult hoatzin is about the size of a chicken. It has a small head, bright red eyes, a feathery comb, and a long tail. Unlike most birds, it has eyes fringed with eyelashes.

Amazing but TRUE

Young hoatzins often stay with their parents for more than a year. They help build new nests and care for their younger brothers and sisters. As many as eight birds may live together in these extended families.

A good crop
Most birds eat insects, worms, and fruits, but the hoatzin eats only the leaves and flowers of certain plants. These leaves are very hard to digest. But that's no problem for this amazing bird. The hoatzin has a special pouch in its throat, called a *crop*, where the leaves are finely ground before being swallowed.

Stinky!
The hoatzin has a strange and unpleasant odor. People who live in the Amazon rainforest have nicknamed it "stinking pheasant."

Tell Me More
Hoatzin chicks can swim and dive from the moment they are born. To protect their young, hoatzin parents build their nest in the fork of a tree branch that hangs over water. When a chick is frightened by a snake or other enemy, it just drops out of its nest into the water. After the danger has passed, the chick swims to shore and climbs back into its nest.

Why are mandrills so colorful?

The female mandrill is not very colorful, but the adult male has so many colors that he is often called the clown of the jungle. He has a long red nose and cheeks that are blue, purple, and white. His beard is orange. The shaggy hair on the upper part of this strange baboon's body is bluish-green. His underparts are tinged with yellow. But that's not all. Both males and females have a brightly colored bottom.

Mandrills spend most of the day searching for food along the ground in the West African rainforest. They travel together in large groups called *troops*. By keeping the male's bright face in sight at all times and following each other's bright bottoms, members of a troop can find each other and stay together, even in the darkest jungle.

Tell Me More

Baby mandrills that are too small to follow the troop grip their mother's underside and hold on for the ride.

Temper! Temper!

When the male mandrill gets excited or angry, his colors turn even brighter. This makes his face look ferocious and scares enemies away.

Calling all mandrills

Mandrills use three different calls to communicate with each other. One is the sound that mothers make when they call to their young. Another is a shriek that the females and the young use when they need help. The third is a rallying cry the male uses to gather the troop. Mandrills also have a special grunt to show they are feeling fine.

Thumbs up

Baboons, like other monkeys and apes, are *primates*. They have thumbs that help them hold objects. Some mandrills pick up seeds, collect fruits and mushrooms, and lift stones to find insects.

Why is the gavial's snout so long?

The Indian gavial is a reptile. It is related to the alligator and crocodile. Like them, the gavial has a long, scaly body and legs that turn out. But the gavial has something that neither of its cousins has: a long, narrow snout that is perfectly shaped for snapping up fish. With more than a hundred razor-sharp teeth in its snout, the gavial can snare a fish with one quick swoop.

Tell Me More

The gavial's eyes, ears, and nose are located on the top of its head. That way it can see, hear, and breathe while the rest of its body remains safely hidden underwater.

Amazing but TRUE

During breeding season, the male gavial grows a big, pot-shaped lump on the end of its nose that the female finds attractive. When an Indian scientist first saw this lump, he named the animal *gharial*, which means "pot" in the Hindi language. The name was later misspelled and passed on as *gavial*.

Why do some moths have "eyes" on their wings?

No moth has real eyes on its wings, but some do have markings, called *eye spots*, that look like eyes. These eye spots are designed to fool and frighten the moth's enemies.

Moths and butterflies have many ways of protecting themselves. Some have shapes and colors that blend in with their surroundings. This makes them almost impossible to see. Others smell or taste bad, which discourages their enemies from eating them for lunch. But the most amazing tricksters are the moths with patterns on their wings that look just like great, staring eyes.

How do eye spots protect a moth?
Whenever it is threatened by a bird or snake, a moth that has eye spots will suddenly spread its wings, flashing two—or even four—big, scary "eyes." This frightens the enemy and gives the moth a chance to escape.

Do poison-arrow frogs shoot arrows?

Despite their name, poison-arrow frogs don't shoot arrows. Unlike poisonous snakes, which bite, these tiny tree frogs don't use their poison to attack others at all. The poison-arrow frog's poison is in its skin and is meant to protect it from predators. A monkey, bird, or snake that decides to eat one of these tiny frogs soon regrets it and is unlikely to make the same mistake again.

Tree babies

Some poison-arrow frogs hide their newly hatched babies, or *tadpoles*, in the rain-filled pockets of flowers that grow on tropical trees. Carrying her tadpoles, the mother frog climbs a tree and places one or two tadpoles into the tiny pool at the center of each flower. She remembers where she leaves her babies and returns to see them from time to time. During these visits, she lays unfertilized eggs, which the tadpoles eat.

Fair warning

Poison-arrow frogs are among the brightest and most colorful creatures in the animal kingdom. Their brilliant colors are a warning to all. They say: POISON! DO NOT TOUCH! STAY AWAY!

Amazing but TRUE

The poison-arrow frog's poison is so strong that it can paralyze a large monkey almost instantly. Indians of the Amazon rainforest rub the poison on the tips of their arrows before they go hunting.

Is the okapi a donkey or a zebra?

The velvety okapi has long ears that look like donkey ears and striped legs that look like zebra legs. Until about a hundred years ago, scientists didn't even know that this beautiful and gentle animal existed. Imagine how surprised they were when they first saw the horse-sized creature in the African rainforest.

Today we know that the okapi is neither a donkey nor a zebra. It is the giraffe's only living relative. Although the okapi doesn't have its cousin's long, long neck, it does have some of the giraffe's other features.

How is the okapi like a giraffe?

Like a giraffe, the okapi male has knobby, skin-covered horns. It also has the giraffe's long black tongue. Scientists believe that the tongue's dark color may help protect it from becoming burned in the hot tropical sun.

Tell Me More

Each okapi has its own special pattern of stripes that is different from all others.

Where's Mommy?

For the first few days of its life, the newborn okapi is left all alone, lying hidden in the forest. The mother goes to it only when it calls.

What's for dinner?

In addition to its main diet of leaves, the okapi eats the burnt wood of trees that have been struck by lightning. It also enjoys euphorbia plants, which are poisonous to human beings.

Amazing but TRUE

The okapi cleans almost every part of its body with its tongue, including its eyes and ears. This tongue is so long, the okapi can even lick its feet from a standing position.

Where are the horseshoe crab's eyes?

The horseshoe crab's tough brown shell, which is more than a foot long, completely covers and protects its soft body and ten legs. All that can be seen outside its shell are two pairs of beady black eyes and a tail. The smaller pair of eyes is on top of the horseshoe crab's back. Two larger eyes peer out from its sides. Using its four eyes, the animal searches for shellfish, worms, and plants along the murky sea bottom.

Moonstruck

In the spring, when the moon is full or new, horseshoe crabs creep onto the beach to mate. The female digs a hole in the sand and lays her eggs in it. The eggs are safe there because right after a new or full moon, the tide recedes. Water will not be able to reach the nest for at least two weeks, when the tide starts to build again.

Tell Me More

Horseshoe crabs live in the shallow waters of the Atlantic Ocean and off the coast of Japan. These hardy animals have managed to survive in almost their original form for about three hundred million years. In fact, they were crawling on the beach long before the dinosaurs roamed the earth.

The horseshoe crab is not a crab. Its only living relative is the spider, but the horseshoe crab doesn't look anything like a spider. It has a long, pointy tail, and its horseshoe-shaped shell is unique in the animal world.

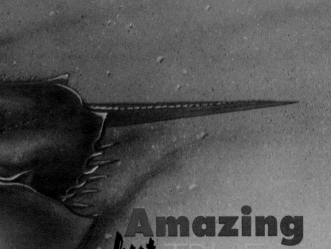

Amazing but TRUE

The horseshoe crab's tail is so sharp and strong that some American Indians used its tip for their arrows.

Does the star-nosed mole's nose shine?

The twenty-two wiggling pink feelers, or *tentacles*, arranged like a star around the star-nosed mole's nose make it different from any other animal on earth. The "star" does not shine at night. But it does help the mole, which can hardly see, find its way in the dark. The star-nosed mole depends on its sensitive tentacles to feel for worms, snails, shrimps, and insects in underground burrows and swamps.

Tell Me More

Star-nosed moles are found in many parts of North America. They swim through the water with ease, using their paddle-shaped legs and long tail. While digging their underground tunnels, the moles keep their tentacles folded over their nostrils to keep out the dirt.

Does the devilfish have horns?

What appear to be two horns on the devilfish's head are really flaps of skin that are used to push food into the animal's mouth. And despite its name, this fish is quite harmless. In fact, devilfish are so friendly that they let deep-sea divers stroke their winglike fins and back.

Does the devilfish actually fly underwater?
The devilfish, also known as the manta ray, looks like a huge bat or bird as it moves gracefully along the bottom of the sea. What appear to be gigantic wings are really fins flapping up and down as the fish swims through the water.

Deep-sea giant
The devilfish is the largest of the rays. It measures about twenty-five feet across from fin to fin—about the length of a school bus. The animal weighs up to two tons.

No bones about it
The devilfish doesn't have a bone in its body. Its skeleton is made of cartilage, the same flexible material that human ears are made of.

Belly flop
Sometimes mantas burst through the surface of the water in spectacular leaps. When they come down again, they hit the water with a big bang.

Tell Me More
Not all rays are as harmless as the manta. Some have poisonous tails. Others, like the sting ray, will give an electric shock if touched on their wings.

Amazing but TRUE
The female devilfish can give birth while leaping out of the water. The live baby fish are actually born in the air! As soon as they emerge from their mother's body, they spread their little winglike fins and dive smoothly into the water.

Why does the hornbill sometimes hide in a hole?

Rhinoceros hornbills are gorgeous birds that live in the rainforests of Asia. Like other hornbills, rhinoceros hornbills have unusual nesting habits. Here is how they raise their family:

1. The female goes inside the hollow of a tree while the male remains outside, making little balls of mud.

2. The female uses the mud to plaster herself into the hole, leaving a small opening.

4. The male gathers food for himself and his mate. When she is hungry, the female sticks her beak out through the opening and the male passes food to her.

5. Weeks later, the chicks hatch. The mother bird comes out at that time. With the help of her mate, she chips away at the mud wall until there is an opening large enough for her to get through. The chicks remain inside their nest and help rebuild the wall, leaving a narrow opening in it.

3 The female lays her eggs inside the hole, where they will be protected from monkeys, snakes, and other egg snatchers.

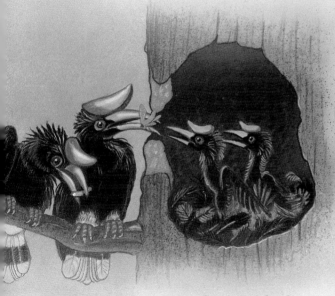

6 Now both parents gather food and feed their chicks through the opening until the young birds are ready to leave the nest.

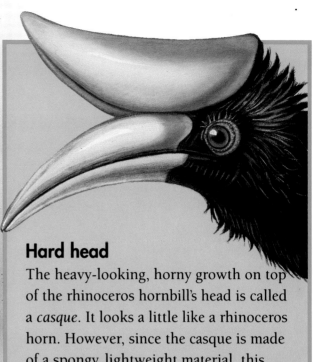

Hard head
The heavy-looking, horny growth on top of the rhinoceros hornbill's head is called a *casque*. It looks a little like a rhinoceros horn. However, since the casque is made of a spongy, lightweight material, this horn is not as heavy as it appears.

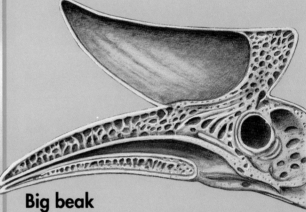

Big beak
Like its casque, this bird's huge "lobster claw" beak is made of light, airy material. If the beak were solid, it would be so heavy that the bird couldn't fly.

Amazing but TRUE

While nesting inside the tree, the female loses all her feathers. But she grows new ones before she comes out again. Mama bird looks fine after her rest, but Papa is thin and worn out from gathering food for the family for so long.

Are armadillos reptiles?

Because of their tough, bumpy skin, armadillos are often confused with turtles or crocodiles. But armadillos are true mammals. They have hair, they give birth to live young, and the females produce milk for their babies.

How did the armadillo get its name?

The armadillo has tough skin made of rows, or bands, of bony plates that look like a coat of armor. Armadillos can be found in the western United States, Mexico, and South America. When the people of Spanish-speaking countries first saw the strange little creature, they named it *armadillo*, or "little armored one."

How tough is the armadillo's armor?

The armadillo's skin is strong enough to protect the animal from the teeth and claws of most of its enemies. But the armadillo is neither fierce nor dangerous. Instead of fighting, this peaceful creature spends most of its time sleeping in underground burrows.

Tell Me More

There are twenty different kinds of armadillos. The giant armadillo is the largest, measuring three feet in length when fully grown. The fairy armadillo is the smallest. It is just a few inches long. Most other armadillos are identified by the number of bands in their armor. Some have as many as twenty-two bands. Most armadillos have three, six, or nine bands.

Play ball

To protect its soft underbelly, the three-banded armadillo can roll itself into a tight, hard ball. Its head and tail are shaped like triangles that fit together like a puzzle.

Busy body

Despite its rigid armor, the armadillo can do some surprising things:

It can swim, dog-paddle fashion.

It can rear up on its hind legs and sniff for danger.

It can anchor itself to the ground with its strong claws.

It can capture sixty to seventy ants with one lick of its long tongue.

Some kinds can walk underwater and stay submerged for up to six minutes.

It can jump up to startle an enemy.

Amazing but TRUE

The nine-banded armadillo always gives birth to quadruplets—four identical babies. There are two kinds of armadillos that give birth to twenty identical babies!

How does the warthog use its tusks?

The warthog is a wild pig that lives on the grasslands of Africa. Its strong, upturned tusks are perfect for digging up the roots and bulbs that the warthog loves to eat. Although warthogs are usually peaceful animals, they can be fierce fighters. When attacked, they use their sharp tusks to defend themselves.

Before their real tusks grow in, young warthogs have fringes of white bristles that look a little like tusks.

Tails up!
The warthog holds its long tail straight up like a flag when it is fleeing from danger. Any other warthog that sees this tail above the tall grass knows that an enemy is near.

Does the warthog really have warts?
Yes. The warthog has three pairs of cone-shaped warts on the sides of its large, flat face. These growths protect the animal's eyes and cheeks as it searches for shoots under thornbushes and other prickly plants.

Babirusa
The babirusa, a wild pig about the same size as the warthog, has longer legs and no warts. Its large head looks even stranger than the warthog's, with two sets of long tusks. One set grows right up through the animal's snout; the other sticks out of its mouth.

Does the matamata have worms on its head?

The matamata turtle has growths below its mouth that look like wriggling worms. These "worms" attract fish that frequently become the matamata's dinner.

When the matamata is hungry, it waits quietly in the murky waters of the Amazon River. If a fish comes close trying to get the "worms," the turtle just opens its mouth and sucks in its prey.

Now you see it, now you don't!

The matamata is a master of disguise. The bumps and ridges on its brownish top shell and its snorkel-like nose blend perfectly with the leaves, roots, and woody stems on the river bottom. When the turtle sits motionless in the muddy water, it is almost impossible to see.

Amazing but TRUE

The matamata's neck is almost twice as long as its backbone. If this were true for humans, it would mean that the average person would have a neck six feet long!

Does the wrasse doctorfish make house calls?

The wrasse is sometimes known as the doctorfish because it helps keep other fish clean and in good health. This brightly colored brave little fish is no more than two or three inches long. Yet it isn't afraid to swim near the largest and most dangerous fish in the Pacific Ocean. Although the wrasse doesn't make "house calls," it does have long "office hours." All the big fish in the area come to its coral-reef "clinic," where they wait patiently for their turn to be treated.

Say "Ahhhh."
Using its thick lips and strong teeth, the wrasse removes parasites and bacteria from many fish every day. First it cleans their eyes and skin. And when it's time to clean their teeth, the large fish willingly open wide. The wrasse then swims inside and does its work.

Mutual benefits

Doctors are usually paid for their services, and the wrasse is no exception. The parasites and bacteria that are harmful to the large fish are nourishment for the wrasse. If the wrasse didn't nibble the parasites off its "patients," it might die of starvation.

Good night, sweet wrasse

After a long day at the "office," the wrasse goes to sleep. When a wrasse rests, it lies on its side and snuggles under the sand. There it can remain safe and undisturbed.

Tell Me More

Although you have come to the last page of this book, you are only beginning to find out about the many unusual members of the animal kingdom. Animals in the wild must adapt to their environment and overcome the danger of predators, starvation, and a harsh climate. The process of continuous adaptation, called *evolution*, can result in some very strange creatures. Many look peculiar, others behave in unexpected ways, and still others have amazing and mysterious powers.

There seems to be a plan and a purpose for everything in nature. Each animal has a role to fulfill. Each has an effect on something else that sooner or later has an effect on us.

Here are some more amazing-but-true facts about unusual animals to start you on your way to new discoveries:

- Certain marine iguanas look like fiery dragons because they puff steam through their nose.

- A barnacle is born with one eye. As it matures, it grows two more eyes. When the barnacle is fully grown, it sheds these two eyes. The one eye that is left then splits in two.

- The horned toad squirts blood from its eyes to frighten its enemies.

- The sphinx moth caterpillar can puff itself up to look like a snake.

- The archerfish shoots down insects with a spray of water from up to three feet away.

- The female walkingstick insect lays one egg at a time, dropping it on the forest floor. At egg-laying time, when thousands of insects lay their eggs, it sounds just like rain.

INDEX

aquatic mammals, 8
archerfish, 32
armadillos, 26–27
    fairy, 27
    giant, 27
    nine-banded, 27
    three-banded, 27
babirusas, 28
baboons, 12–13
barnacles, 32
butterflies, 15
cassowaries, 6–7
devilfish, 22–23
doctorfish, 30–31
gavials, 14
hoatzins, 10–11
    crops, 11
hornbills, 24–25
horned toads, 32
horseshoe crabs, 20
manatees, 8–9
manta rays, 22–23
mandrills, 12–13
    troops, 12
marine iguanas, 32
matamatas, 29
mermaids, 8
moths, 15
    eye spots, 15
    sphinx, 32
okapis, 18–19
platypuses, 4–5
poison-arrow frogs, 16–17
    tadpoles, 16
primates, 13
rhinoceros hornbills, 24–2
    casques, 25
sea cows, 8–9
    manatees, 8–9
    Steller's, 8
star-nosed moles, 21
    tentacles, 21
walkingsticks, 32
warthogs, 28
wrasses, 30–31